THE LIFEBOAT COLLECTION

AN ILLUSTRATED GUIDE TO THE HISTORIC LIFEBOAT COLLECTION AT CHATHAM DOCKYARD

NICHOLAS LEACH AND PETER WOOLHOUSE
A LIFEBOAT ENTHUSIASTS' SOCIETY PUBLICATION

▲ The enthusiastic volunteers looking after the Historic Lifeboat Collection are always pleased to welcome vsitors to Chatham Historic Dockyard. (Tony Buckingham, by courtesy of the RNLI)

ISBN 978-1-909540255
© Nicholas Leach and Peter Woolhouse 2022
The rights of Nicholas Leach and Peter Woolhouse to be identified as the authors of this work have been asserted in accordance with the Copyrights, Designs and Patents Act 1988. All rights reserved. No part of this book may be reprinted or reproduced or utilised in any form or by any electronic, mechanical or other means, now known or hereafter invented, including photocopying and recording, or in any information storage or retrieval system, without permission in writing from the publishers.

Published by
Foxglove Publishing/Lifeboat Enthusiasts' Society
Foxglove House, Shute Hill, Lichfield, Staffs WS13 8DB
t > 07940 905046
e > njl7@outlook.com

Layout and design by Nicholas Leach/Foxglove Publishing

This is a not an official RNLI publication. Any views or opinions expressed in this publication are solely those of the authors and do not necessarily represent those of the RNLI.

CONTENTS

FOREWORD 5
By Mark Dowie, RNLI Chief Executive

THE LIFEBOAT COLLECTION 6
The Lifeboat Fund 8
The move to Chatham 9
The volunteer group 13

THE LIFEBOATS 15
St Paul 16
Lizzie Porter 17
Louisa Heartwell 18
James Leath 20
BASP 21
Helen Blake 24
Susan Ashley 25
St Cybi (Civil Service No.9) 26
North Foreland (Civil Service No.11) 28
Grace Darling 29
J.G. Graves of Sheffield 30
The Will & Fanny Kirby 31
44-001 32
Edward Bridges (Civil Service No.37) 34
Spirit of Lowestoft 36
A-504 38
B-500 39
D-112 40
Georgina Taylor (D-727) 41
Olive Laura Deare (E-002) 42
Lifeguard craft 43
Lifeboats in the collection (summary) 44

Foreword

I am pleased to be able to introduce this guide to the Historic Lifeboat Collection, the UK's largest collection of historic lifeboats. Opened to the public in 1996, the Collection presents the RNLI's history through a variety of rescue craft, information boards, models and a number of other artefacts.

No fewer than 20 life-saving craft are preserved and tell the story of the rescues, the crews and the courage behind almost 200 years of gallant service. The values of service, selflessness and dependability are still very much at the heart of today's RNLI, with the charity marking its bicentenary in 2024.

In the best traditions of the organisation, the Collection is cared for and maintained by a team of dedicated volunteers, who devote much time and effort to conserving our heritage.

I hope you enjoy visiting the Historic Lifeboat Collection, and that finding out more about the lifeboats in this book will enhance your experience.

Mark Dowie
Chief Executive
Royal National Lifeboat Institution

THE LIFEBOAT COLLECTION

The RNLI Historic Lifeboat Collection consists of twenty lifeboats and numerous artifacts, which tell the story of rescue work around the United Kingdom and Ireland over the past two centuries. The collection encompasses pulling and sailing lifeboats of the last century, through to the high-speed inshore lifeboats of the 1960s to the modern 54ft Arun and 47ft Tyne class lifeboats. The lifeboats range from the virtually 'as-found' state of the 1897-built St Paul to the more modern boats, which retain their in-service equipment and appearance.

The Collection also covers the history of the men and women who served aboard the boats, and there are displays of self-righting lifeboats and different hull shapes for lifeboats. There are launching carriages, a launching tractor, engines, winches and other equipment of past and more recent times which were all important elements of the shore support so vital at the RNLI's lifeboat stations. Housed in the No.4 Covered Slip at the Historic Dockyard Chatham, in Kent, the collection is a unique record of life-saving around the coasts of the United Kingdom and Ireland.

The Historic Dockyard has proved to be an ideal home for the Lifeboat Collection, which in 2021 marked a quarter of a century since its opening, and forms part of a wider display of maritime history at the Dockyard.

▼ Helen Blake is a core part of the Historic Lifeboat Collection.

◀ The dockside general cargo shed, known formerly as 'L Shed', at Bristol City Docks which was home to the National Lifeboat Museum for a decade from 1981. It adjoined the City's own Industrial Museum on Prince's Wharf.

It forms one attraction of the many on offer at the Historic Dockyard, being visited by those who might not otherwise have considered visiting a lifeboat display. The boats are under cover and protected from the elements in a well ventilated building, sufficiently open to the atmosphere to prevent the kind of drying out and shrinkage which might result from being kept in an enclosed building.

The idea of a national lifeboat museum began in the late 1970s and early 1980s. The RNLI has always been conscious of its heritage, yet its charter prevented it from spending money raised for saving life at sea on presenting its heritage to the public. However, it retained a number of historic boats which had been withdrawn from service, and some had been initially loaned to a privately-run museum at Bristol, where they were joined by others owned by the museum or lent by individual owners. The Museum leased the ground floor

◀ The hull of B.A.S.P. being lifted into the Bristol Museum. The boat was in private ownership for twenty-five years before coming to Bristol, and has been gradually restored by volunteers at both Bristol and Chatham.

THE LIFEBOAT COLLECTION 7

▲ Lizzie Porter (left) and Helen Blake (right) in situ at the Bristol Museum.

of a disused dockside cargo shed from Bristol City Council adjoining the City's Industrial Museum on Prince's Wharf. The City was developing the area for leisure purposes as Bristol City Docks were no longer being used by commercial shipping.

One of the more unusual exhibits was the 28ft lifeboat *Helen Blake*, the only example of its type, which was owned privately but loaned to the Museum, arriving in Bristol on 1 November 1984. Before going on display, she had been fully overhauled and repainted. The refurbishment programme lasted nearly eighteen months and was carried out by apprentices of Fairey Marine Ltd, the company who succeeded her builder, Groves & Guttridge, at Cowes. There were still staff at the yard who were involved in her original construction more than forty years earlier. After this, she was transported to Bristol and presented to the Museum at a civic reception held on 12 April 1985 in the presence of the Lord Mayor of Bristol, Councillor Claude Draper.

Unfortunately, despite the best efforts of those involved and the acquisition of many interesting exhibits, the Bristol museum ran

THE LIFEBOAT FUND

The names of some lifeboats in the historic collection include 'Civil Service No...'. This is a reference to a fund which was set up over 150 years ago by Civil Servants (who then included post office employees) to raise money to fund the building of new lifeboats. It was originally known as the Civil Service Life-Boat Fund, but this was subsequently shortened to The Lifeboat Fund, which is itself a registered charity.

Since its establishment, the Fund has provided 53 lifeboats for the RNLI and Lifeboat Pier on the River Thames. The Lifeboat Fund has raised over £26 million for the RNLI, providing not only new lifeboats but also providing equipment and training. There are currently nine lifeboats in RNLI service paid for by the Fund.

▲ The lifeboat Civil Service No.6 (ON.384) served at Douglas from 1896 to 1924.

▲ (left) North Foreland on display at Bristol; (right) Susan Ashley and D-148 installed in the Bristol museum.

into financial difficulties and closed. However, the boats remained on site and under cover, although not accessible to the public and in need of a new home. But when Bristol City Council wanted their building back, extra impetus was given to the problem of finding a suitable new home for the collection.

The move to Chatham

The RNLI's Lifeboat Preservation Working Group had existed for almost twenty-five years, finding homes for individual historic boats. One of its members, Simon Stephens of the National Maritime Museum, was aware of the space available at The Historic Dockyard's eighty-acre site in Chatham. Although several towns had expressed interest in housing the Collection, none had a suitable building ready, and the finance would always have been a problem. However, the Historic Dockyard (itself a charity) was a different proposition, for among its large number of listed buildings was the enormous and under-used No.4 Covered Slip, a 146-year-old building of architectural significance.

The infrastructure needed for a major display was also in place, such as ticket offices, toilets, refreshments, parking and so on. The Dockyard would store the boats but, as it preferred them to be on public display,

◀ St Cybi (Civil Service No.9) was displayed at the Scottish Maritime Museum in Irvine for almost ten years before being moved to Chatham.

▶ Grace Darling arrived at Bristol on 28 October 1985, directly out of RNLI service, for permanent exhibition.

▼ Volunteers helping with the Lifeboat Collection, pictured in the mid-1990s, take a break from cleaning, polishing and painting the historic craft in their care.

a partnership was soon developed. The Dockyard was keen to add another attraction to offer visitors and the RNLI would be able to develop a National Collection of lifeboats. So, in the spring of 1994 nine lifeboats and two inshore lifeboats were transported from Bristol to Chatham, and placed in Slip No.4, on gravel, forming a cosmetic shingle 'beach', paid for by the Lifeboat Enthusiasts' Society (LBES). However, the collection needed formal management, proper displays and information boards.

The funding of such displays was beyond the available resources of either party, but a successful application to the the National Lottery's Heritage Fund saw a grant of £355,000, topped up by another £100,000 from a private trust fund, being made. Although the major work was contracted out to commercial concerns, the employment of staff to care for the collection was out of the question, so enlisting the help of volunteers was an obvious solution. Meanwhile, a few more lifeboats joined the collection in various states of repair, and plans were eventually formulated for the creation of a permanent museum.

The Historic Lifeboat Collection (HLC) Volunteer Group was established, and the members, who have cared for the collection since the mid-1990s, soon had the boats looking presentable. On Bank Holiday Saturday, 25 May 1996, the long-held hopes of the RNLI came to fruition and the unique National Collection of Lifeboats opened its doors to the public for the first time. A formal opening by the Duke of Kent, President of the RNLI, took place on 11 September 1996.

In 1997 the prototype 44ft Waveney, 44-001, straight out of the RNLI's relief fleet, arrived at the Dockyard and was initially kept afloat. She was moored on the River Medway adjacent to the Dockyard, and the Volunteer Group undertook maintenance, including running the engines and later taking her out for 'exercise' runs on the river. With the approval of the RNLI management, 44-001 attended events on the Thames to promote the RNLI, and participated in the RNLI's 175th Anniversary celebrations at Poole in 1999.

Afloat activities came to an end in 2000, when 44-001 was lifted out of the water to become a static exhibit in the Collection. However, a syndicate of four volunteers had the opportunity to acquire another former lifeboat, the Rother class *Mary Gabriel*, which became part of the HLC with the agreement of the RNLI, until 2016.

▲ E-002 Olive Laura Deare arriving at Chatham in 2013.

▼ The 37ft 6in Rother *Mary Gabriel* moored at Thunderbolt Pier, June 2003. (Steve Renyard)

THE LIFEBOAT COLLECTION

▲ The 47ft Tyne class lifeboat Spirit of Lowestoft in situ at Chatham.

Significant additions to the HLC were not made for many years, but on 6 August 2013 one of the first E class lifeboats built for Thames service, E-002 *Olive Laura Deare*, was formally handed into the care of the HLC team. The E class boat was followed in 2019 by the 47ft Tyne *Spirit of Lowestoft*, one of the last of its class in service and another lifeboat involved in a medal-winning rescue. In the same year, the Arancia inshore rescue boat A-03, modified to become a flood rescue boat, arrived. This boat was notable having been used in a Bronze Medal-winning service at Umberleigh, Devon in December 2012.

Another addition came in 2021 when the pulling and sailing lifeboat *Louisa Heartwell* was acquired by the RNLI. This significant lifeboat had been involved in a Gold

▶ The Fowler tractor T65 served at Gourdon, Hastings and Dungeness and represents launching vehicles within the Collection. Tractors have played a major role in getting beach-launched lifeboats afloat since the 1920s, when they replaced horses.

medal-winning rescue off Cromer with Coxswain Henry Blogg at the helm, and she was deemed a significant craft and thus worthy of preservation despite having been in private ownership for a number of years. She was donated to the RNLI during 2020 and kept at the All-weather Lifeboat Centre at Poole until a suitable opportunity to transfer her could be found, but this was delayed by the Covid pandemic and its lockdowns.

The volunteer group

A few months after the lifeboat collection was moved to Chatham, a meeting was held at the Dockyard on 15 October 1994, chaired by Ray Kipling, then Deputy Director of the RNLI, and attended by John Francis, founder and honorary secretary of the LBES, with other members of the Society and local RNLI branch members. An outline of the aspirations for the collection and future plans were presented, and volunteers at the meeting were asked to give whatever time they could.

The outcome of that meeting was the formation of a group of around twenty active volunteers, who have worked tirelessly to maintain the RNLI's collection. Of this group, six have been there since that first meeting in 1994, seven have sadly passed away and others have left, but

▼ The Chatham volunteers with the 37ft Oakley lifeboat The Will and Fanny Kirby in August 2018. On the boat, left to right: Pete Birthright, Ken Sandercock, David Cox and Lance Wright; standing, left to right: Keith Parrott, Ken Sherlock, Chris Cooke, Ian Smith (Gallery Manager), Ann Reed, Steve White, Nick Berry (Deputy Manager), Trevor Murless, Paul Severns, Ian Russell (Deputy Manager), Tina Smith, John Turrell, Peter Woolhouse and Steve Prescott. (Tony Buckingham, by courtesy of the RNLI)

▲ HRH The Duke of Kent visited Chatham in July 2022.

▼ Volunteer Anne cleaning one of the information boards. (Tony Buckingham, by courtesy of the RNLI)

new volunteers have come forward to ensure a steady core of people who continue to clean, paint and restore the boats. Conditions for the volunteer group were basic in the early days, and tea breaks were taken in the open, even through the first winter. Later, a 'portacabin' crew room was installed, and this is still in use.

The volunteer group has had only three 'managers', although the title of the role has changed three times. Steve Renyard was the first, and he worked hard to establish the foundations of the Historic Collection, being in charge from 1994 to 2005. He was succeeded by Peter Dawes, who headed the group from 2005 until 2016. Peter was followed by Ian Smith, the current manager, who is also a full-time crew member at the Gravesend lifeboat station, one of the four stations on the Thames.

The volunteer group has continued to maintain the fine collection of historic boats at minimal cost to the RNLI, as funds available for heritage projects are limited. Conservation and restoration projects have also been undertaken, with the latest project being to conserve the former Cromer lifeboat *Louisa Heartwell*.

As members of the group get older, it is to be hoped that new volunteers will continue to join to keep the group strong. New volunteers are always welcome – no special skills required!

The Lifeboats

St Paul	page 16
Lizzie Porter	17
Louisa Heartwell	18
James Leath	20
BASP	21
Helen Blake	24
Susan Ashley	25
St Cybi (Civil Service No.9)	26
North Foreland (Civil Service No.11)	28
Grace Darling	29
J.G. Graves of Sheffield	30
The Will & Fanny Kirby	31
44-001	32
Edward Bridges (C.S. & P.O. No.37)	34
Spirit of Lowestoft	36
A-504	38
B-500	39
D-112	40
Georgina Taylor (D-727)	41
Olive Laura Deare (E-002)	42
Lifeguard craft	43

THE LIFEBOAT COLLECTION

St Paul

NAME	St Paul
OFFICIAL NUMBER	406
BUILT	1897, James Beeching, Great Yarmouth
DIMENSIONS	38ft x 12ft
TYPE	Norfolk & Suffolk pulling & sailing, 12 oars
STATIONED	Kessingland
RECORD	13 launches, 18 saved

▼ St Paul being hauled out of the lifeboat house at Kessingland. (By courtesy of the RNLI)

▶ After service, St Paul was converted into the yacht Stormcock in private hands.

The clinker-built *St Paul* is the oldest lifeboat in the Lifeboat Collection. She was built in 1897 by the well-known Great Yarmouth boatbuilder James Beeching, and served at Kessingland in Suffolk for thirty-four years, launching thirteen times on service during that time. She served as the penultimate lifeboat at the station, which was closed in 1936.

In 1919 she was involved in a fine service, launching on the afternoon of 11 December to the sailing smack *A.J.W.*, of Rye, which had gone aground on the Newcombe Sands in rough seas. Getting *St Paul* afloat through the heavy surf pounding the beach was difficult, but with skill and effort the lifeboat got clear. To effect a rescue, the lifeboat had to be anchored and veered down to the sunken smack, enabling the lifeboat crew to rescue four survivors. For this rescue Silver Medals were awarded to Coxswain George Knights and Second Coxswain Edward Smith.

After being withdrawn from service, *St Paul* was sold out of service in 1931 and became the yacht *Stormcock*. Her long-term care at Chatham is under review.

Lizzie Porter

The pulling and sailing self-righter *Lizzie Porter* is a good example of the design of lifeboat typical of RNLI operations for much of the late nineteenth and early twentieth centuries. Funded from the legacy of Miss E. Porter of Halifax, she was placed on station at Holy Island, in Northumberland, in October 1909. During more than fifteen years there, she saved seventy-seven lives.

Her most notable rescue took place on 15-16 January 1922, when she saved nine men from the trawler *James B. Graham*, which was ashore on rocks. The whole village turned out to help launch the lifeboat in the dark and snow. For this rescue the Silver Medal was awarded to Coxswain George Cromarty.

She was transferred to North Sunderland in July 1925, when a motor lifeboat replaced her at Holy Island, and was sold out of service in 1936. Little is known about what happened to her over the next forty years, but she was discovered derelict in the River Trent in the 1970s.

A request for assistance was answered by the Army, who removed the rotting vessel from where she had lain, partially submerged, for some years. The double diagonal planking was badly perforated, but she was taken to the National Lifeboat Museum at Bristol, where she was fully restored.

NAME Lizzie Porter
OFFICIAL NUMBER 597
BUILT 1909, Thames Iron Works, Blackwall
DIMENSIONS 35ft x 8ft 6in
TYPE Self-righter, ten-oared
STATIONED Holy Island; North Sunderland
RECORD 56 launches, 77 saved

▼ Lizzie Porter under oars while serving at North Sunderland. (By courtesy of the RNLI)

LOUISA HEARTWELL

NAME	Louisa Heartwell
OFFICIAL NUMBER	495
BUILT	1902, Thames Iron Works, Blackwall
DIMENSIONS	38ft x 10ft 9in
TYPE	Liverpool pulling and sailing, 14 oars
STATIONED	Cromer
RECORD	115 launches, 195 saved

▼ Louisa Heartwell on her carriage ready to be launched from the beach at Cromer, with a large crowd watching. Note the Tipping's Plates fitted to the wheels of the carriage. These were specially developed for stations where carriage launching was employed and were intended to prevent the wheels from sinking into soft sand.

Louisa Heartwell, a non-self-righting Liverpool class pulling and sailing lifeboat, was built in 1902 at a cost of £981 12s, which was largely paid for by a £700 legacy. She was propelled either by sails or fourteen oars. She was the sixth lifeboat to be stationed at Cromer, on the Norfolk coast, and served there for twenty-nine years.

Her second coxswain was the renowned Henry Blogg, who became the RNLI's most decorated lifeboatman, being awarded three Gold Medals, four Silver Medals, the George Cross, BEM, and a number of other awards during his career. He was awarded the first of his Gold Medals for a service undertaken on Louisa Heartwell in January 1917, when he led his crew in assisting the Swedish steamer Fernebo, after an explosion had broken the vessel in two, in gale force conditions.

Louisa Heartwell, which had only just returned from a previous service, was relaunched off the beach through heavy surf, with the assistance of bystanders. Once she had successfully got away, she was able to rescue eleven survivors. In addition to the Gold Medal awarded to Blogg, Sliver Medals went to the rest of the crew for their courage and endurance during the two demanding services.

Louisa Heartwell

◀ Louisa Heartwell shortly after she arrived at Chatham. The deck and cabin were added during her time private ownership, and were removed soon after she was acquired by the Collection.

▼ The initial work on Louisa Heartwell at Chatham involved the removal of all non-lifeboat related structures added during her post-service life. She will undergo a significant conservation, with the aim of returning her as close to her original lifeboat condition as is possible, so she becomes the boat commanded by Henry Blogg.

Louisa Heartwell was sold out of RNLI service in 1931, and was converted to a single screw ketch with an engine. She was kept at Barry Dock and also Brancaster. She was later converted to a houseboat, and was in use on the Grand Union Canal, moving to the Chichester Canal, Sussex, in 1997.

Louisa Heartwell is one of only seven surviving pulling and sailing Liverpool class lifeboats of the forty built by the RNLI. She was offered back to the RNLI for restoration after becoming the property of Premier Marinas. Her notable history made her worthy of preservation, and she was moved to the RNLI's All Weather Lifeboat Centre in Poole before being transported to the Lifeboat Collection at the Dockyard in spring 2021.

James Leath

NAME	James Leath
OFFICIAL NUMBER	607
BUILT	1910, Thames Iron Works, Blackwall
DIMENSIONS	42ft x 12ft 6in
TYPE	Norfolk & Suffolk, 12-oared
STATIONED	Pakefield; Caister; Aldeburgh
RECORD	32 launches, 20 saved

▼ James Leath, weighing more than eight tons, is manoeuvred on the beach at Pakefield, being dragged across the shingle beach with skids laid down to make the process easier.

The Norfolk & Suffolk type lifeboat, of which James Leath is a fine example, was primarily designed for sailing, being strongly built with a heavy iron keel that made them extremely stable. The high decks enclosed watertight compartments to make them buoyant, although they were not self-righting.

James Leath was built in 1910 and was stationed at Pakefield, Caister and Aldeburgh, thus serving in both Suffolk and Norfolk. At the stations she served, she was one of two lifeboats; it was common practice during the pre-1914 era for No.1 and No.2 lifeboats to be operated from East Anglian stations.

In December 1926, while she was on station at Caister, James Leath was nearly lost to the sea when a spring tide and strong north-westerly winds swept away part of the beach, and the two lifeboats which were kept on the beach had to be hauled over 100 yards uphill to safety.

Sold out of service in August 1935, James Leath was used as a houseboat at Poole. A large wheelhouse and petrol engine driving a single screw were fitted. In 1972 she was in use as the yacht *Robin Hood II* owned by a Mr Marsden, of Hamworthy, Poole and kept in Holes Bay. Restoration work on the boat began when she was at the Bristol Museum, was continued at Chatham, and remains ongoing.

B. A. S. P.

This 45ft Watson motor lifeboat *B.A.S.P.* was built in 1924 and her unusual name was an amalgamation of the legacies used to fund her: Mrs M.P. Smart, Richard Blackburn, Mrs H.D. Price and Mrs Constance Armstrong. She was one of the RNLI's early motor lifeboats, and was powered by a single 80hp White petrol engine.

Her service career, which lasted more than thirty years, began at Yarmouth on the Isle of Wight in October 1924, where she served for ten years. She next spent six years at Falmouth, from 7 November 1934, after which she was placed in the reserve fleet in January 1940. She served in a temporary capacity at stations during the Second World War including Penlee, Rosslare, Ballycotton, Dunmore East, Arranmore and Dun Laoghaire.

In May 1947 she took up duty at Valentia in south-west Ireland after the station had been reopened by the RNLI. She stayed there until January 1951, when was returned to the reserve fleet. A further four years as a reserve lifeboat took her to Donaghadee, Arranmore, Howth and Dun Laoghaire.

The much-travelled lifeboat was sold out of RNLI service in February 1955 for £550 and was used as a work boat for almost thirty years. She was rescued from a mud berth in 1978 and taken to the nascent National Lifeboat Museum in Bristol in the early 1980s to be restored.

NAME	B.A.S.P.
OFFICIAL NUMBER	687
BUILT	1924, J S White, Cowes
DIMENSIONS	45ft x 12ft 6in
TYPE	Watson motor
STATIONED	Yarmouth; Falmouth; Reserve; Valentia
RECORD	86 launches, 37 saved

▲ B.A.S.P. on duty at Yarmouth, the first station from which she operated. (By courtesy of the RNLI)

THE LIFEBOAT COLLECTION

HELEN BLAKE

NAME	Helen Blake
OFFICIAL NUMBER	809
BUILT	1938, Groves & Guttridge, Cowes
DIMENSIONS	28ft x 8ft
TYPE	Harbour, single screw
STATIONED	Poolbeg
RECORD	13 launches, 5 saved

▲ Helen Blake served as Poolbeg lifeboat, covering Dublin Bay.

Helen Blake, sole example of the 28ft Harbour class of motor lifeboat, was an interesting development of 1938 which foreshadowed the need at certain stations for a smaller type of rescue boat. Had the war not intervened it is possible that the class would have multiplied.

Poolbeg station, intended to cover work only inside the Liffey estuary, had a light 30ft non-self-righting whaleboat type pulling lifeboat, and there was no comparable class of motor lifeboat with which to replace her. The Harbour class was designed by Richard Oakley, later to become the RNLI's surveyor of lifeboats, and *Helen Blake* was built in 1938. With a hull divided into eight watertight compartments and fitted with twenty-nine air cases, she was unusual for a lifeboat in that she had a transom stern.

Helen Blake served at Poolbeg from 1938 to 31 October 1959, when the station was closed. She was sold out of service to RNLI Secretary Lt Col Charles Earle, and with few alterations became the yacht *Sea Call* before she was overhauled in the late 1970s at Cowes to be put on display.

THE LIFEBOAT COLLECTION

Susan Ashley

The 41ft Watson motor lifeboat *Susan Ashley* was built shortly after the end of the Second World War and was sent to Sennen Cove, Cornwall in July 1948. Her aluminium alloy superstructure was an innovation at the time, being a shelter over the engine room providing the crew a little protection from the worst of the weather. She was fitted with a radio and loud hailer, and her twin Weyburn AE6 six-cylinder petrol engines, each of 35bhp, gave her a top speed of 7.82 knots.

Provided from the legacy of Charles Carr Ashley, who died at Mentone in France in 1906, she was named after the donor's mother, being the third lifeboat in the RNLI fleet to be so named. As marine technology advanced, small marine diesel engines were developed and were introduced into the lifeboat fleet. Many existing lifeboats were re-engined including *Susan Ashley*, which was fitted with twin 47hp Parsons Porbeagle diesels in 1963.

After spending twenty-five years at Sennen, *Susan Ashley* was replaced in May 1973 and spent six years at Barry Dock as the No.2 lifeboat, launching from the slipway there, before being used as a boarding boat at Tynemouth for just over a year.

NAME	Susan Ashley
OFFICIAL NUMBER	856
BUILT	1948, Groves & Guttridge, Cowes
DIMENSIONS	41ft x 11ft 8in
TYPE	Watson motor, twin screw
STATIONED	Sennen Cove; Barry Dock No.2; Tynemouth (BB)
RECORD	94 launches, 67 saved

▼ Susan Ashley being recovered up the slipway at Sennen Cove.

THE LIFEBOAT COLLECTION

St Cybi (C.S. No.9)

NAME	St Cybi (Civil Service No.9)
OFFICIAL NUMBER	884
BUILT	1950, J.S. White, Cowes
DIMENSIONS	52ft x 13ft 6in
TYPE	Barnett, twin screw
STATIONED	Holyhead; Relief
RECORD	257 launches, 161 saved

▶ St Cybi (Civil Service No.9) on trials in July 1950. She was the second of her type to be built. (Beken, by courtesy of the RNLI)

▼ St Cybi (Civil Service No.9) afloat in Holyhead harbour, after having radar and a self-righting air-bag fitted, and having had her cockpit enclosed. (By courtesy of Holyhead Maritime Museum)

One of twenty 52ft Barnetts built, *St Cybi (Civil Service No.9)* spent almost her entire RNLI career at Holyhead, saving more than 150 lives during her time on Anglesey. She was named at the station on 16 June 1951 by Lady Hopkins, wife of the Right Hon Sir Richard V. N. Hopkins, chairman of the Civil Service Lifeboat Fund, which had funded the boat. She was built with an open steering position which was later enclosed to form a wheelhouse.

The most famous rescue in which she was involved took place on 2 December 1966 when, together with the Moelfre lifeboat *Watkin Williams*, she went to the aid of the Greek motor vessel *Nafsiporos*, which was in distress in a hurricane force north-westerly wind and very rough seas. *St Cybi* and the Holyhead crew rescued five from the motor vessel, with the Gold Medal being awarded to Lt Cdr Harold Harvey, Inspector of Lifeboats, who had taken the wheel at the Coxswain's request. Coxswain Richard Evans from Moelfre was also awarded the Gold Medal for this service.

Silver Medals were awarded to Holyhead Coxswain Thomas Alcock and Motor Mechanic E.S. Jones, with Bronze Medals going to

St Cybi (C.S. No.9)

the rest of the crew for their bravery during this service, which is one of the most famous in RNLI history.

St Cybi was involved in two further medal-winning services. On 3-4 September 1971 she went to the aid of the yacht Sinbad. Four men were taken off, three of whom were unconscious as a result of carbon monoxide poisoning. Bronze Medals were awarded to Mechanic Donald Forrest and Gareth Jones, and a Bronze Second Service Clasp to John Hughes.

The other outstanding service took place on 11 September 1976, when the lifeboat launched to the yacht Pastime in distress twenty-three miles south-west of Skerries lighthouse. Using only one engine, Coxswain William Jones made an approach and rescued the four crew, for which he was subsequently awarded the Silver Medal.

St Cybi (Civil Service No.9) served at Holyhead for thirty years, enjoying an illustrious career, during which time she launched a total of 257 times and saved 161 lives. She was taken out of service in 1986 and placed on display at the Scottish Maritime Museum, Irvine, being moved to Chatham in 1995.

▲ Lt Cdr Harold Harvey was District Inspector of Lifeboats. His involvement in the Nafsiporos rescue turned him into one of the most famous RNLI employees in the history of the service.

NORTH FORELAND

NAME	North Foreland (Civil Service No.11)
OFFICIAL NUMBER	888
BUILT	1951, J.S. White, Cowes
DIMENSIONS	46ft 9in x 12ft 9in
TYPE	Watson, twin screw
STATIONED	Margate; Relief
RECORD	395 launches, 216 saved

▼ The scene during the naming ceremony of North Foreland at Margate on 17 May 1951.

A total of twenty-eight Watsons of 46ft 9in in length were built between 1947 and 1957, and *North Foreland (Civil Service No.11)* was one of three built in 1951. She was named on 17 May 1951 at Margate by HRH The Duchess of Kent. During her service career, she was fitted with a wheelhouse to enclose her midships steering position and an airbag was installed on her aft cabin in the 1970s to give her a once-only righting capability.

North Foreland served in the RNLI fleet for three decades, of which 27 years saw her stationed at Margate, where she was kept in a boathouse on the pier and launched from a slipway. She launched on service at Margate 389 times and saved more than 200 lives.

The most noteworthy rescue in which *North Foreland* was involved took place on 7 November 1952 when she and the Margate crew went to the aid of the auxiliary barge *Vera*, which had gone aground in the Thames Estuary. She reached the casualty in the early hours of the morning, having battled through severe weather, to save two men who were clinging to the rigging; they had been there for five hours. Coxswain Denis Price was awarded the Silver Medal for this outstanding rescue.

GRACE DARLING

Provided out of the RNLI's general funds and named after the lighthouse keeper's daughter involved in the famous rescue in 1838, *Grace Darling* was notable for being the last 35ft 6in Liverpool class lifeboat to be built. In April 1954 she was sent to North Sunderland, the lifeboat station closest to the Farne Islands where her namesake's renowned exploits took place.

The most noteworthy rescue she carried out was on 12 July 1959, in south-westerly force seven to eight winds and rough seas, when she went to Inner Farne to rescue a canoeist in difficulty. For this rescue, the Bronze Medal was awarded to Coxswain Thomas Dawson.

Grace Darling served in the RNLI fleet for three decades, from 1954 to 1984, during which time she was re-engined, her original twin 20hp Ferry FKR3 engines being replaced in 1971 by slightly more powerful 32hp Parsons Porbeagle diesels.

After four years in the relief fleet, which included a spell at Flamborough, in 1971 she was sent to Youghal, on the south coast of Ireland, where she was launched down a small slipway into the sheltered River Blackwater.

NAME	Grace Darling
OFFICIAL NUMBER	927
BUILT	1954, Groves & Guttridge, Cowes
DIMENSIONS	35ft 6in x 10ft 8in
TYPE	Liverpool motor, twin screw
STATIONED	North Sunderland; Relief; Youghal
RECORD	106 launches, 34 saved

▲ Grace Darling on trials just after she had been built.

J. G. Graves of Sheffield

NAME	J.G. Graves of Sheffield
OFFICIAL NUMBER	942
BUILT	1958, William Osborne, Littlehampton
DIMENSIONS	37ft x 11ft 6in
TYPE	Oakley, twin screw
STATIONED	Scarborough; Relief; Clogher Head; Newcastle
RECORD	147 launches, 21 saved

▼ J.G. Graves of Sheffield on trials prior to entering service.

The 37ft Oakley class was a ground-breaking design when it was introduced in the late 1950s, with *J.G. Graves of Sheffield* being the first boat of the new class. The boat was designed by and named after Richard Oakley, who realised that making a lifeboat self-righting was possible using a method which involved the transfer of 1.54 tons of water ballast within the hull.

Hitherto, self-righting lifeboats were narrower in the beam than non-self-righters and thus more likely to capsize, but the new 37ft boat incorporated the best features of both self-righting and non-self-righting boats.

The new design had many benefits, including its weight of between nine and eleven tons making it suitable for handling ashore and carriage launching over a beach, yet the water ballast taken on once afloat provided good sea-keeping qualities.

J.G. Graves of Sheffield, the first of twenty-six 37ft Oakley lifeboats to be built, was sent to Scarborough in October 1958, and after twenty years of service there was transferred to the relief fleet. She later served at Clogher Head and Newcastle in Ireland, leaving service in 1993 after an impressive thirty-five year career.

The Will and Fanny Kirby

The *Will and Fanny Kirby* was the fifth 37ft Oakley to enter service, and was built for Seaham as a replacement for the lifeboat *George Elmy*, which capsized on service in November 1962 with the loss of all five of the lifeboat crew and four of the five people they had taken off the casualty. Powered by twin 43hp Perkins P4M engines, she was re-engined in 1982 with more powerful 52hp Thorneycroft diesels.

At Seaham, she is credited with saving sixty-six lives. On 11 November 1973 she was involved in a service to anglers trapped on the north pier, saving eighteen people, for which the Bronze Medal was awarded to Coxswain Arthur Farrington. When she was withdrawn from Seaham on 24 February 1979, the station was closed.

She then entered the relief fleet, and when on duty at Hoylake, she was involved in another fine service, saving the catamaran *Truganini*, of Mostyn, and her three exhausted crew in a force nine westerly wind, gusting to storm force ten, and very rough seas on 20 September 1979. After this service. the Bronze Medal was awarded to Coxswain Thomas Jones and the Thanks of the Institution Inscribed on Vellum accorded to David Dodd.

NAME	The Will and Fanny Kirby
OFFICIAL NUMBER	972
BUILT	1963, William Osborne, Littlehampton
DIMENSIONS	37ft x 11ft 6in
TYPE	Oakley, twin screw
STATIONED	Seaham, Relief, Flamborough
RECORD	197 launches, 116 saved

▼ The Will and Fanny Kirby on station at Seaham.

44-001

OPERATIONAL NUMBER 44-001

BUILT 1964, USCG Yard, Baltimore

DIMENSIONS 44ft 10in x 12ft 8in

TYPE Waveney, twin screw

STATIONED Trials and Relief Fleet; Falmouth

RECORD 311 launches, 109 saved

▶ 44-001 on trials, with her wheelhouse painted white. She spent three years on trials, between 1964 and 1967. (By courtesy of the RNLI)

▼ 44-001, with a crew of Chatham volunteers, taking part in the RNLI's 175th anniversary celebrations in Poole in 1999.

The historically significant 44ft Waveney class 44-001 was the first 'fast' lifeboat to see service with the RNLI. Developed and built in America, the United States Coast Guard 44ft motor lifeboat revolutionised British lifeboat design when it was introduced in the 1960s. The design was showcased at the International Lifeboat Conference in Edinburgh in 1963, after which the RNLI acquired a new boat from the USCG, number 44328, for trials around the United Kingdom and Ireland.

The US boat, given the number 44-001 by the RNLI, was taken on an extensive tour of the British Isles, visiting many lifeboat stations so that crews' opinions of the new design could be gauged. The reaction of crews who saw the boat was so positive that a building programme was initiated. Given the class name Waveney, a total of twenty-one more boats were built in Britain for the RNLI.

After trials, 44-001 was used as a relief lifeboat for thirty years, serving at many stations. The most notable rescue in which she was involved took place at Eyemouth on 6 October 1990, when she launched at

44-001

night in hurricane force winds and 35ft seas to rescue two skin divers who were in danger near rocks. The storms had closed Eyemouth harbour, but the lifeboat put to sea despite the conditions, and for this service the Silver Medal was awarded to Acting Coxswain James Dougal.

44-001 was taken out of service in 1996, and was used as a floating exhibit at Chatham Historic Dockyard for four years. In June 1999, crewed by Chatham volunteers, 44-001 participated in the celebrations at Poole, to mark the RNLI's 175th Anniversary. On 8 May 2001 she was taken out of water and has since been displayed ashore as a static exhibit.

▲ 44-001 attending Kingston-upon-Thames RNLI Regatta 1998 crewed by volunteers from the Historic Lifeboat Collection.

▼ 44-001 on display at the entrance to the Historic Lifeboat Collection at Chatham.

THE LIFEBOAT COLLECTION

Edward Bridges

NAME Edward Bridges (Civil Service No.37)

OFFICIAL NUMBER 1037

BUILT 1974, Wm Osborne, Littlehampton

DIMENSIONS 54ft x 17ft

TYPE Arun, twin screw

STATIONED Torbay

RECORD 456 launches, 285 saved

▶ Edward Bridges (Civil Service No.37) at moorings in Brixham Harbour. She was named at Brixham on 17 June 1975 by HRH The Duke of Kent. (By courtesy of the RNLI)

▼ Acting Coxswain Keith Bower of Torbay. He was presented with the Gold Medal for conspicuous gallantry for rescuing ten people from the cargo vessel Lyrma in force ten winds, increasing to force eleven. During the rescue, the lifeboat encountered seas of up to 40ft head-on. (By courtesy of the RNLI)

The 54ft Arun class *Edward Bridges (Civil Service No.37)* was the third of the class, and was the first of seven 54ft Aruns to be constructed; all the other Aruns were 52ft in length. She was the third and last Arun to be built of wood, and was completed by William Osborne in October 1974, after which she went on evaluation trials.

Power came from two 460hp Caterpillar D343 engines, and her cost of £162,383 was met by the Civil Service Lifeboat Fund. She was named after the late Baron Bridges, a distinguished civil servant.

She was placed on service at Torbay on 16 April 1975 and went on to serve the station for almost two decades, gaining an outstanding record of service, being credited with saving almost 300 lives. The most outstanding service in which she was involved took place on 6 December 1976, when she saved ten people from the cargo vessel *Lyrma*, of Panama, for which Acting Coxswain Keith Bower was awarded the Gold Medal while Bronze Medals went to the rest of the crew.

Another noteworthy service by *Edward Bridges* took place on 19 February 1978, when she launched to the pilot cutter *Leslie H*. At one point the lifeboat was knocked down by a giant wave, estimated at more than 30ft in height, and a crew member was swept overboard, but was immediately recovered. Three men were then taken off the cutter, which later sank. The Bronze Medal was awarded to Coxswain George Dyer for this service and medal service certificates went to the crew.

Edward Bridges

Towards the end of the same year, on 2 December, another challenging service was undertaken when six crew were saved from the trawler *Fairway* in storm force winds and heavy seas. Coxswain Arthur Curnow manoeuvred the lifeboat alongside the casualty twice to effect a rescue and, for his skill, he was awarded the Bronze Medal.

In April 1994 *Edward Bridges* was taken to Berthon Boat Co at Lymington for a survey, which revealed severe deterioration in the boat's hull and decking and, as a result, the lifeboat was withdrawn from service. She remained at Berthon's yard for more than a year, until a decision was made to place her on display at Chatham.

▲ Edward Bridges (Civil Service & Post Office No.37) is the largest lifeboat in the Lifeboat Collection.

▼ Edward Bridges was one of three wooden-hulled Aruns.

THE LIFEBOAT COLLECTION

Spirit of Lowestoft

NAME	Spirit of Lowestoft
OFFICIAL NUMBER	1132
BUILT	1987, Fairey Allday Marine, Cowes
DIMENSIONS	47ft x 15ft
TYPE	Tyne, twin screw
STATIONED	Lowestoft
RECORD	560 launches, 83 saved

▶ Spirit of Lowestoft at Lowestoft in 2009. She served the station with distinction for twenty-seven years before coming to Chatham.

▼ The 47ft Tyne was designed for stations where slipway launching was employed, such as Baltimore in south-west Ireland, where Hilda Jarrett is pictured launching on exercise. She was built at the same time as Spirit of Lowestoft.

The 47ft Tyne lifeboat *Spirit of Lowestoft* was built in 1987 for Lowestoft in Suffolk. She was the twenty-second Tyne class boat, a design developed during the late 1970s, with the first boat entering service in 1982. *Spirit of Lowestoft* cost £520,166, and arrived at Lowestoft in November 1987. She spent twenty-seven years at the Suffolk station, gaining a fine record of service there.

The Tyne class was designed as a fast replacement lifeboat for stations where slipway launching was employed, although several were kept afloat, including *Spirit of Lowestoft*. In order to fit into existing lifeboat houses, the Tyne had a low profile, and the distance from the keel to the top of the wheelhouse had to be within 13ft.

The most noteworthy service by *Spirit of Lowestoft* took place in storm force winds on 29 August 1996, when she launched, along with Aldeburgh lifeboat *Freddie Cooper*, to the yacht *Red House Lugger*, which was in difficulty about thirty miles south-east of Lowestoft. The lifeboats managed to evacuate the people on board the yacht, which was then towed to Harwich, in a long service

SPIRIT OF LOWESTOFT

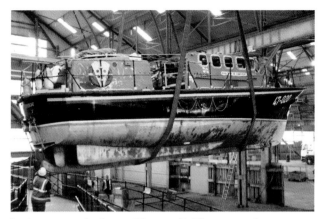

◀ Spirit of Lowestoft arriving at Chatham in June 2019 ready to join the Historic Lifeboat Collection (HLC) at Chatham.

lasting over twelve hours. The coxswains of both lifeboats were awarded the RNLI Bronze Medal.

Spirit of Lowestoft left Lowestoft in October 2014 and, after four years in the relief fleet, was taken out of service. Instead of being sold, a decision was made to preserve her as part of the RNLI's heritage, and she was added to the Historic Lifeboat Collection in June 2019.

▼ Spirit of Lowestoft at the Historic Lifeboat Collection. She was powered by twin 425hp GM diesel engines, being re-engined in 1997 with twin 565hp units, making her the most powerful lifeboat in the Chatham collection.

A-504

OFFICIAL NUMBER	A-504 originally 18-004)
BUILT	1970, William Osborne, Littlehampton
DIMENSIONS	18ft 6in x 8ft
TYPE	McLachlan, twin screw
STATIONED	Weston-super-Mare
RECORD	171 launches, 60 saved

▼ A-504 in her launching cradle at the head of the slipway at Weston-super-Mare. (By courtesy of the RNLI)

Designed by J.A. McLachlan, the ragged-chine-hulled 18ft 6in inshore lifeboat was one of several new designs the RNLI considered as they looked for an inshore boat with a greater capability, including night operations, than the standard inflatable, which had entered service a few years earlier.

The unusual ragged chine hull reduced the hull surface in contact with the water and thus minimised the pounding found with normal hard-chine boats. Ten boats to McLachlan's design were built, most (including A-504) of glass reinforced plastic. A-504 was stationed at Weston-super-Mare from May 1970 to May 1983, in conjunction with an ILB, being launched from a cradle which was lowered down the long slipway from the lifeboat house on Birnbeck Island.

She was involved in many rescues, with a particularly noteworthy one taking place on 13 September 1975, when she rescued five men from a motor boat which was stranded on the rocks at Brean Down in a strong north by easterly gale. The Bronze Medal was awarded to Helmsman Julian Morris in recognition of the courage, determination and seamanship he displayed during this service.

B-500

Developed at Atlantic College, at St Donat's Castle in South Wales, the Atlantic 21 was the first rigid-inflatable lifeboat to see service with the RNLI. The first few craft, of which B-500 was one, had in-line seating for the crew, but this was soon modified and the seats were arranged in a delta-shape behind the steering console in the centre of the boat.

The rigid hull supported the weight of the boat and, without the inflatable sponsons in contact with the water, gave the boat good sea-keeping capabilities and a high speed, with the boats achieving more than thirty knots. The hulls were originally built of plywood and divided into watertight compartments, but, when full-scale production began, hulls were moulded from glass reinforced plastic (GRP).

B-500 was used mainly for trials around the coast at several stations. She had a four-year stint in the relief fleet, before being transferred to the Bristol Museum for display in July 1980. The RNLI continued to develop the rigid inflatable, and the current version of the type, the Atlantic 85, is the third generation in the fleet.

OFFICIAL NUMBER	B-500
BUILT	1970
DIMENSIONS	21ft 4in x 7ft 6in
TYPE	Atlantic, twin outboards
STATIONED	Appledore, Hartlepool, Largs, Littlehampton, New Brighton, Queensferry
RECORD	70 launches, 28 saved

▲ The prototype Atlantic 21 B-1 ('Bravo 1') became B-500 and was trialled operationally at Hartlepool and Appledore. (By courtesy of the RNLI)

D-112 Blue Peter III

OFFICIAL NUMBER	D-112
BUILT	1967
DIMENSIONS	15ft 6in x 6ft 4in
TYPE	D class inflatable, outboard
STATIONED	North Berwick, Relief
RECORD	33 launches, 20 saved

▼ D-112 Blue Peter III at North Berwick taking part in a first aid exercise, with a casualty being lifted onto the ILB. (By courtesy of the RNLI)

Blue Peter III (D-112) was one of four D class inflatables funded from an appeal by the BBC TV programme Blue Peter in the 1960s. They were subsequently replaced following further such appeals. Blue Peter-funded lifeboats eventually served at seven stations, six of which had ILBs while a seventh, Fishguard, was provided with a Trent class all-weather lifeboat.

The D class inflatable ILB was introduced in 1963. The RNLI acquired an inflatable boat in 1962 for trials, and a delegation visited France, where similar boats were in operation, to obtain advice and see the boats in action. Following initial testing and trials, the first inshore rescue boats were introduced in the summer of 1963, when eight were sent to the coast. Such was their success that in the following years an increasing number of stations began to operate inflatable boats.

D-112 was stationed at North Berwick from 1967 to 1972 and then entered the relief fleet, serving at Trearddur Bay (1982), Little and Broad Haven (1983), Scarborough (1983) and Margate (1984). She was one of the lifeboats to be displayed at the Bristol Museum.

D-727 GEORGINA TAYLOR

D-727 Georgina Taylor at Tenby during the station's lifeboat day, August 2015.

The D class inflatable inshore lifeboat has been developed and improved by the RNLI since the 1960s, and the latest incarnation is known as the Inshore Boat 1 (IB1). Between 2000 and 2003 the IB1 project saw the RNLI's naval architects re-examine the inflatable design and, as a result, the ILB was completely re-engineered. A faster craft was developed, incorporating the latest advances in material and equipment technology, with a more powerful engine and a higher speed.

D-727 represents the IB1, and is very different from D-112. Named *Georgina Taylor* on 20 March 2010 at Tenby by executor Richard Gray, she was the third inshore lifeboat to be funded for Tenby by Georgina Taylor. She was placed on station on 8 December 2009, and served until October 2021.

OFFICIAL NUMBER	D-727
BUILT	2009
DIMENSIONS	4.95m x 2m
TYPE	D class inflatable, outboard
STATIONED	Tenby

▼ D-727 newly arrived at Chatham Dockyard.

THE LIFEBOAT COLLECTION

E-002 OLIVE LAURA DEARE

OFFICIAL NUMBER E-002	
BUILT 2001, Tiger Marine	
DIMENSIONS 9m x 2.94m	
TYPE E class, twin waterjets	
STATIONED Gravesend (also Tower and Chiswick)	
RECORD 660 launches, 43 saved	

▲ E class Mk.1 lifeboats Olive Laura Deare (E-002) (behind) and Public Servant (Civil Service No.44) (E-001) off Gravesend in the Thames Estuary, 2006.

In January 2001 the RNLI announced that it would be operating lifeboats on the tidal reaches of the River Thames in response to a request from the Maritime and Coastguard Agency to provide a rescue service. To fulfil this undertaking, a new type of lifeboat, the E class, was introduced into service.

The service officially commenced in January 2002 and the E class entered service for exclusive use at the three Thames lifeboat stations of Gravesend, Tower and Chiswick. The Mk.I boats were based on a 9m fast response craft design from Tiger Marine Ltd. The boats were built at FBM Babcock Naval Shipyard, Rosyth, and powered by twin 240hp Steyr 246 marine diesels each driving a Hamilton waterjet.

Six Mk.I E class boats were built, of which *Olive Laura Deare* (E-002) was the second. The individual boats were not permanently assigned to any station, but were operated between the three stations.

The boats were extensively modified in 2003 and 2004, being fitted with an intercom set-up, radar/plotter, two VHF radios, and new navigation lights and deck floodlights. E-002, the first boat to be modified, was redelivered to the Thames in July 2003. She remained in service until 2012.

LIFEGUARD RESCUE CRAFT

◄ RNLI Lifeguards use a variety of equipment, including Personal Water Craft, which can respond quickly to incidents with swimmers off beaches.

The RNLI has operated lifeguard services on beaches around the United Kingdom and Ireland since 2001 following a successful pilot scheme. In order to undertake this service, a variety of craft are used, including PWC (personal water craft) and small Arancia type inflatables, which can be launched quickly from a beach.

▼ The Arancia A-03FR, used by the RNLI's Flood Rescue Team, on display alongside one of the Lifeguards Personal Water Craft, more commonly known as a jet-ski. A Flood Rescue Team used A-03FR to rescue a woman swept from her car near Umberleigh in Devon on 23 December 2012. She was clinging to a tree and surrounded by flood waters. The boat's helm, Martin Blaker-Rowe, was awarded a Bronze Medal for the service, the first such award made to recognise a flood rescue.

Lifeboats in the Collection

ON*	Built	Name	In service	Stations
LIFEBOATS				
406	1897	St Paul	1897 – 1931	Kessingland
495	1902	Louisa Heartwell	1902 – 1931	Cromer
597	1909	Lizzie Porter	1909 – 1936	Holy Island; North Sunderland
607	1910	James Leath	1910 – 1935	Pakefield; Caister; Aldeburgh
687	1924	B.A.S.P.	1924 – 1955	Yarmouth; Falmouth; Valentia; Relief
809	1938	Helen Blake	1938 – 1959	Poolbeg
856	1948	Susan Ashley	1948 – 1981	Sennen Cove; Barry Dock; Tynemouth
884	1950	St Cybi (C.S.No.9)	1950 – 1986	Holyhead; Relief
888	1951	North Foreland (C.S.No.11)	1951 – 1981	Margate; Relief
927	1954	Grace Darling	1954 – 1984	North Sunderland; Relief; Youghal
942	1958	J.G. Graves of Sheffield	1958 – 1994	Scarborough; Relief; Clogher Head
972	1963	The Will & Fanny Kirby	1963 – 1993	Seaham; Relief; Flamborough
–	1964	44-001	1964 – 1996	Relief; Falmouth; Relief
1037	1974	Edward Bridges (C.S.No.37)	1975 – 1994	Torbay
1132	1987	Spirit of Lowestoft	1987 – 2018	Lowestoft; Relief
INSHORE LIFEBOATS				
D-112	1967	Blue Peter III	1967 – 1984	North Berwick; Relief
D-727	2009	Georgina Taylor	2009 – 2021	Tenby
B-500	1970	[Un-named]	1970 – 1977	Hartlepool; Mudeford; Lyme Regis; Littlehampton; Largs; Appledore
A-504	1970	[Un-named]	1970 – 1983	Weston-Super-Mare
E-002	2002	Olive Laura Deare	2002 – 2012	Gravesend (also Tower and Chiswick)
A-03	–	[Un-named]		Flood Rescue
	–	[Un-named]		Personal Water Craft (jet ski) (three)
TRACTOR				
T65	1958	Fowler [reg no. YVT 878]	1958 – 1990	Gourdon; Hastings; Dungeness; Relief

*Since 1884 almost every RNLI all-weather lifeboat has been allocated an Official Number (abbreviated to ON) by which it can be identified. These consecutive numbers give an approximate indication of the year in which the boat was built. Official Numbers are recorded on a plaque mounted inside the lifeboat. The Official Number is different from the Operational Number, which appears on the hull of modern lifeboats (such as 44-001, 47-020, etc).

All photographs are by Nicholas Leach, Peter Woolhouse or supplied by the RNLI, unless otherwise credited.